GEOMETRIC DESIGNS·1
COLORING·BOOK

ARTIST·INFO

ISBN-13:978-1717400086

L·E·N·S TRAFFIC

LUX·ET·NATURA·SECULO

ALL PAGES ARE CENTERED
BETWEEN EDGE AND DOTTED LINE
CUT ALONG DOTTED LINE TO REMOVE

THIS IS YOUR TEST PAGE

TRY-OUT PENCILS, PENS, MARKERS, PAINTS, ETC.

PLACE BLANK PAPER BETWEEN PAGES TO PREVENT BLEED-THROUGH

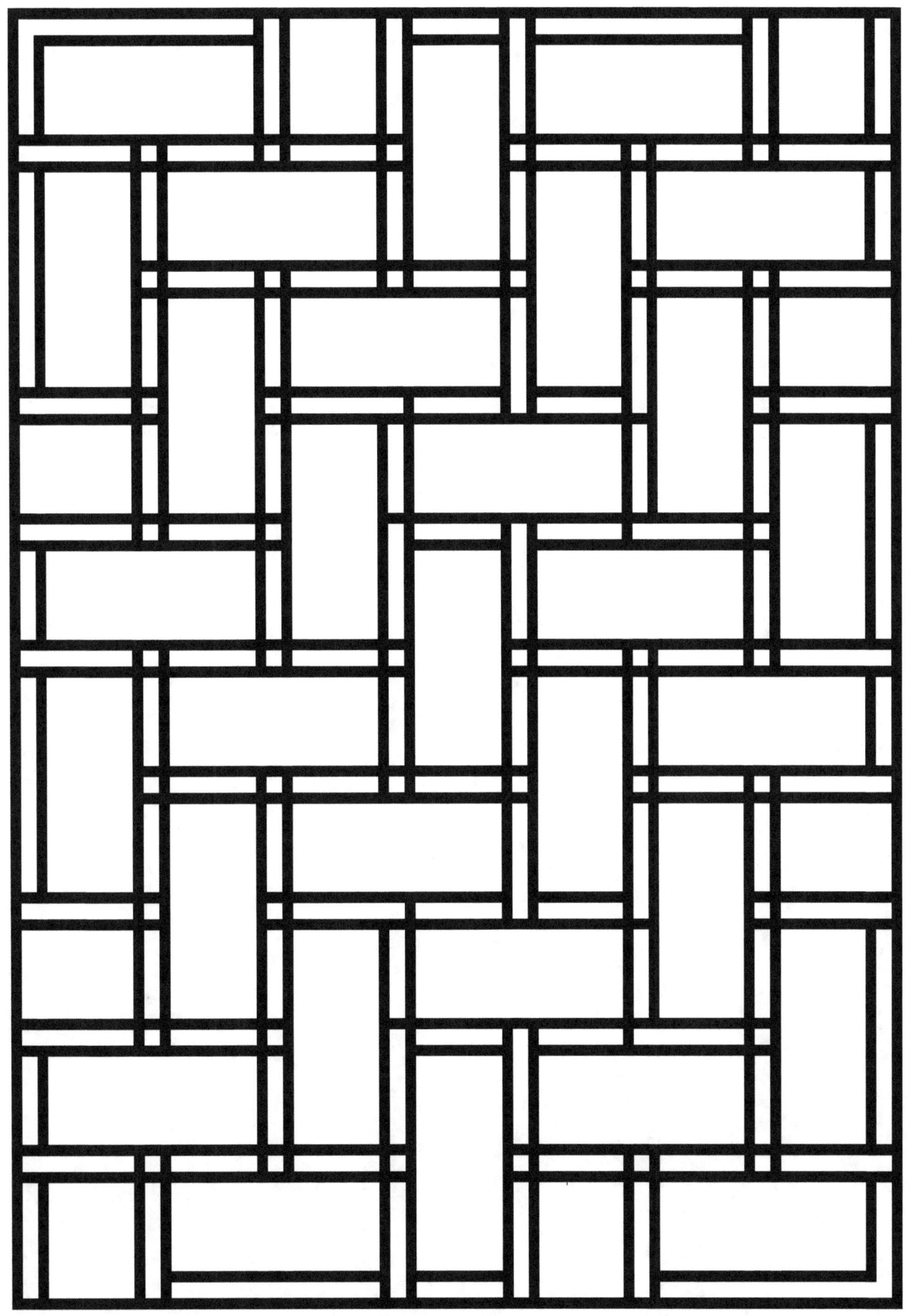

GEOMETRIC·DESIGNS·2

SAMPLE·PAGE

FIBONACCI·DESIGNS·1

SAMPLE·PAGE

IMPOSSIBLE·DESIGNS·1

SAMPLE·PAGE